7

Night is coming in this city.

Can you find the circles that **shine** light on the street?

City Shapes

Circles

By Jan Kottke

Welcome Books

Children's Press
A Division of Grolier Publishing
New York / London / Hong Kong / Sydney
Danbury, Connecticut

Photo Credits: Cover, all photos ©Indexstock.
Contributing Editor: Jennifer Ceaser
Book Design: Michael DeLisio

Visit Children's Press on the Internet at:
http://publishing.grolier.com

Cataloging-in-Publication Data

Kottke, Jan
    Circles / by Jan Kottke.
      p. cm.—(City shapes)
    Includes bibliographical references and index.
    Summary: Observations of circular objects include lighted signs, streetlight globes, traffic lights,
      and clocks.
    ISBN 0-516-23075-1 (lib. bdg.)—ISBN 0-516-23000-x (pbk.)
    1.   Circle—Juvenile literature   [1.   Circle]
I.   Title   II.   Series
    2000
516'.15—dc21

Contents

This place has many shapes.

Can you find the circle that looks like **Earth**?

5

The gumballs in the **machine** are circles.

Name the colors of the circles that you see.

Look at all the **signs**!

Some are squares, some are rectangles, and some are circles.

How many circle signs do you see?

This **fence** has circles all in a row.

How many circles are there?

Do you see the clocks on the sides of the building?

Do you think they are circles?

15

Here are circles that light up.

Which circles tell you to stop?

How many of these circles
can you count?

You can take a bike for a ride.

Do you see any circles that can **spin** around?

Circles can be big, little, flat, and round.

Circles in the city are everywhere.

New Words

Earth (**urth**) one of the nine planets
that move around the sun

fence (**fens**) a metal guard put around
a garden or a field

machine (ma-**sheen**) an object with
moving parts

shine (**shyn**) to give off a bright light

signs (**synz**) things that tell you where
to go

spin (**spin**) turn around

To Find Out More

Books

Circles and Squares Everywhere!
by Max Grover
Harcourt Brace & Company

Circus Shapes
by Stuart J. Murphy and Edward Miller
HarperCollins Publishers

Colors & Shapes
by David A. Carter
Simon & Schuster Children's

Index

About the Author

Jan Kottke is the owner/director of several preschools in the Tidewater area of Virginia. A lifelong early education professional, she is completing a phonics reading series for preschoolers.

Reading Consultants

Kris Flynn, Coordinator, Small School District Literacy, The San Diego County Office of Education

Shelly Forys, Certified Reading Recovery Specialist, W.J. Sahnow Elementary School, Waterloo, IL

Peggy McNamara, Professor, Bank Street College of Education, Reading and Literacy Program